Robots
on the Run, Roll, or Stroll!

Mary Lou Brown & Sandy Mahony

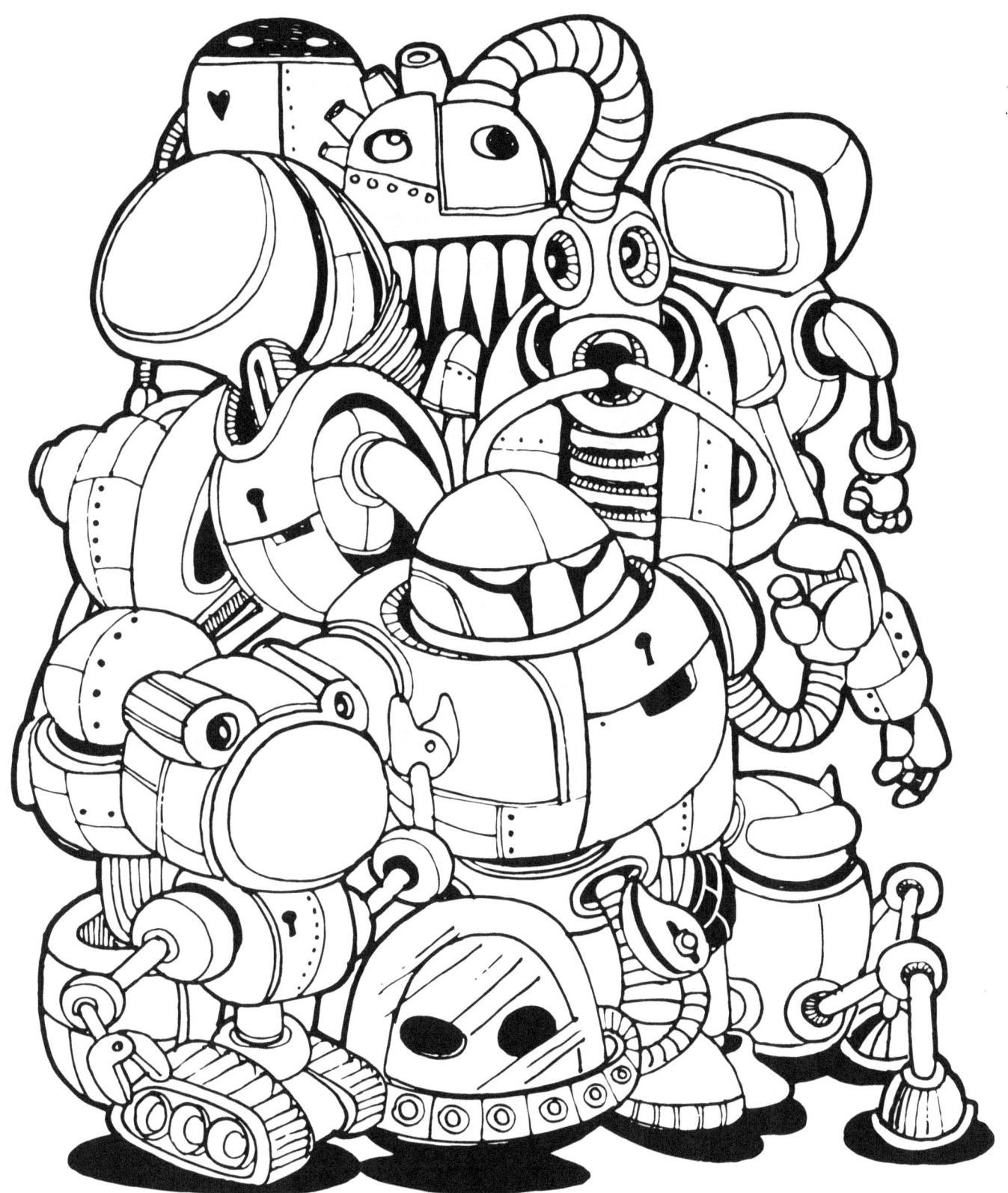

Copyright ©2016 Mary Lou Brown & Sandy Mahony
All rights reserved. No part of this book may be reproduced in any form or by any electronic or mechanical means including information storage and retrieval systems, without permission in writing from the authors. The only exception is by a reviewer, who may quote short excerpts in a review.

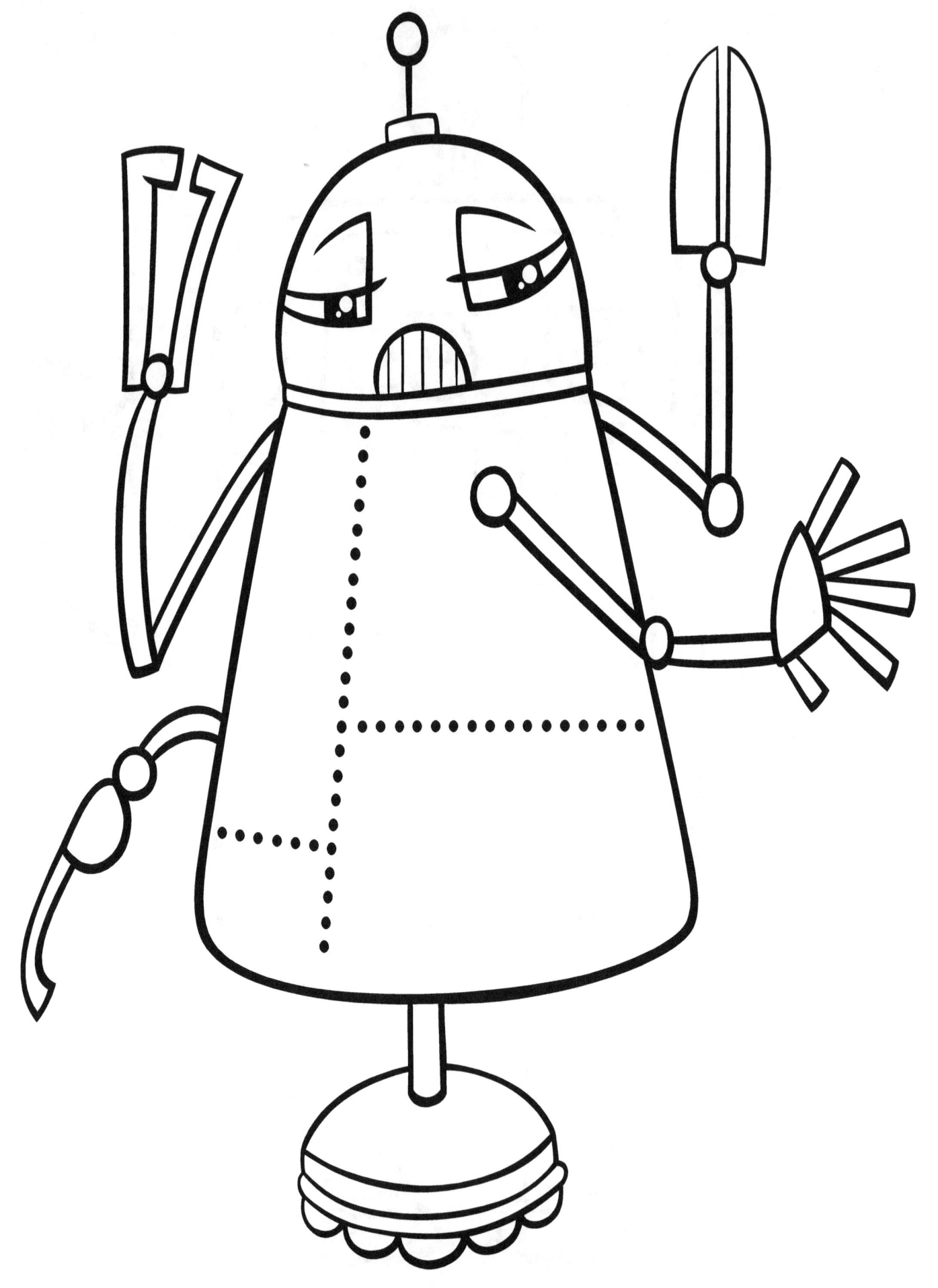

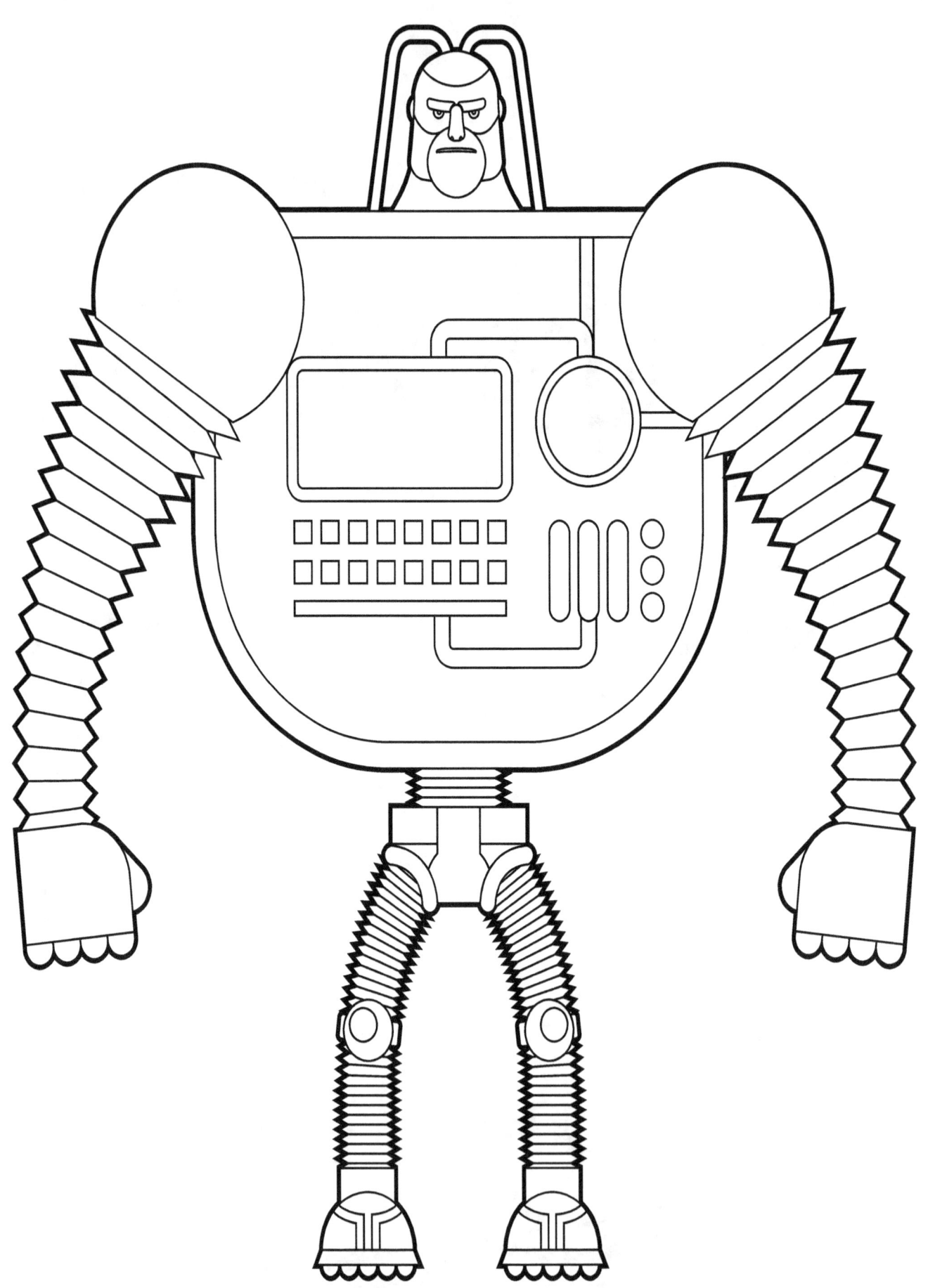

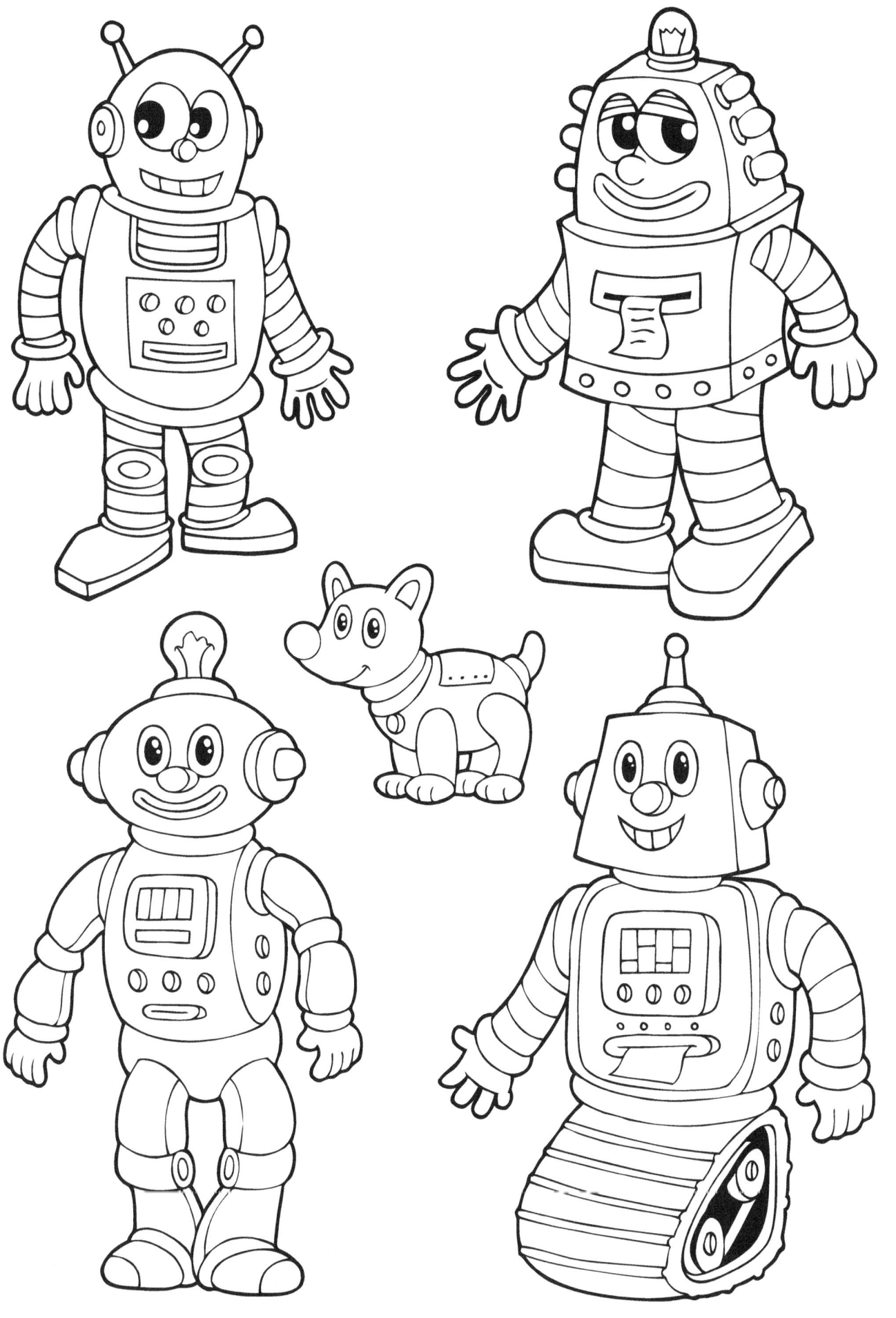

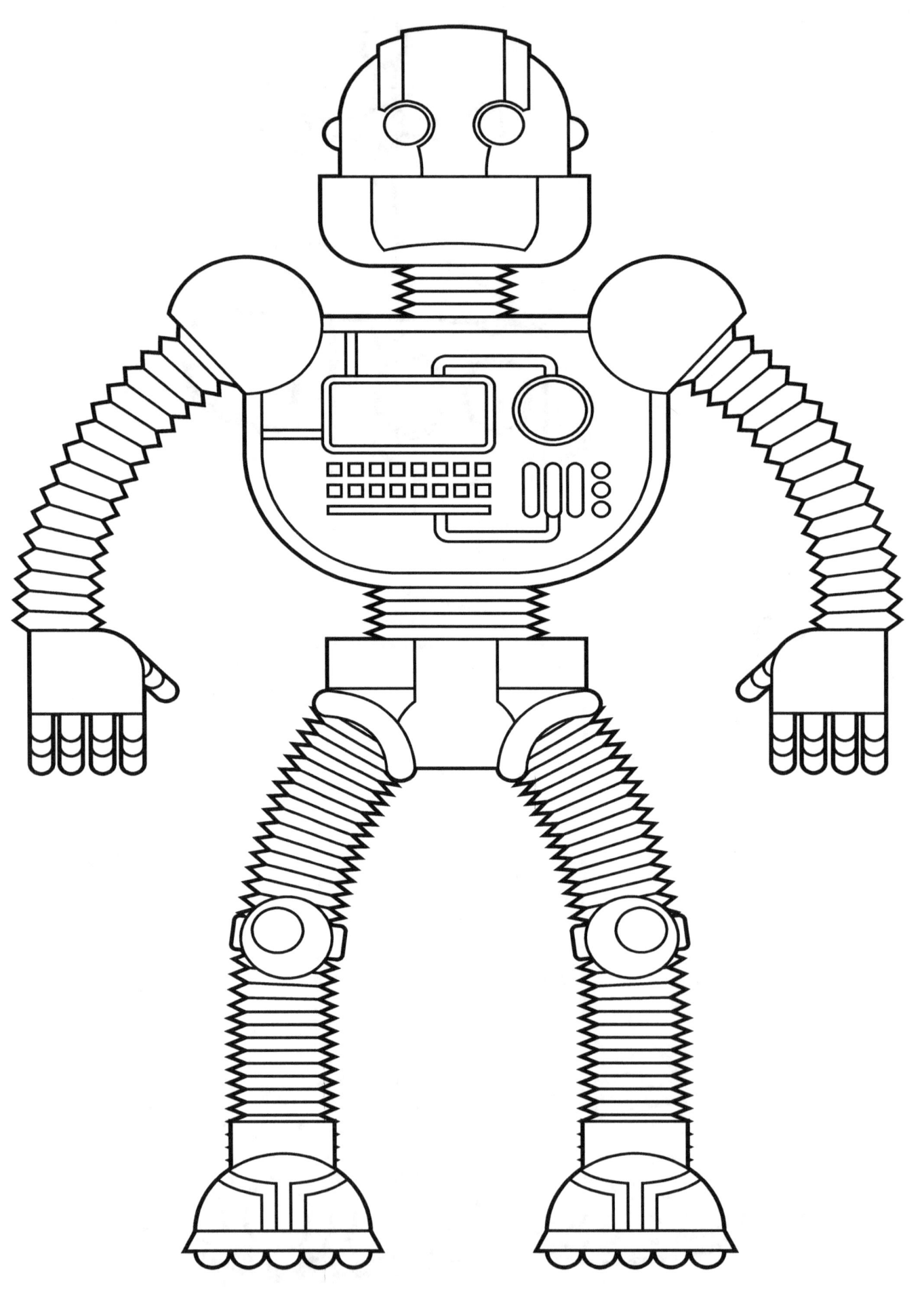

www.ingramcontent.com/pod-product-compliance
Lightning Source LLC
Chambersburg PA
CBHW080524190526
45169CB00008B/3040